A TEXT BOOK OF

TAXONOMY OF ANGIOSPERMS

(PAPER – IV)

FOR

B.Sc. Part-I (Botany) Semester-II

As Per New Revised CBCS Pattern Syllabus of Punyashlok
Ahilyadevi Holkar Solapur University, Solapur
(June 2019)

Dr. D. N. Kutwal
M.Sc., M.Phil, Ph.D.
Ex. Head of Botany Department,
Shankarrao Mohite Mahavidyalaya,
Akluj.

Dr. K. U. Garad
M.Sc., Ph.D.
Assistant Professor,
Head of Botany Department,
Santosh Bhimrao Patil College of Arts,
Commerce & Science, Mandrup.

Prof. A. S. Chandanshive
M.Sc.
Associate Professor,
Department of Botany,
K.B.P. College, Pandharpur.

Prof. R. I. Savalajkar
M.Sc., M.Phil. Associate Professor
Head of Botany Department,
Shankarrao Mohite Mahavidyalaya,
Akluj.

Prof. M. D. Satpute
M.Sc. M.Phill.
Associate Professor
Head of Botany Department,
K.B.P. College, Pandharpur

Dr. N. M. Pise
M.Sc., Ph.D.
Assistant Professor,
Department of Botany,
K.B.P. College, Pandharpur.

NIRALI PRAKASHAN
ADVANCEMENT OF KNOWLEDGE

N5217

B.SC. PART-I: BOTANY (P-IV) (SEM. II)   ISBN 978-93-89686-67-8

First Edition : **December 2019**

© : **Authors**

Published By :
NIRALI PRAKASHAN
Abhyudaya Pragati, 1312, Shivaji Nagar,
Off J.M. Road, PUNE – 411005
Tel - (020) 25512336/37/39, Fax - (020) 25511379
Email : niralipune@pragationline.com

➤ DISTRIBUTION CENTRES

PUNE

Nirali Prakashan : 119, Budhwar Peth, Jogeshwari Mandir Lane,
Pune 411002, Maharashtra. Tel : (020) 2445 2044, 66022708,
Fax : (020) 2445 1538, Email: bookorder@pragationline.com,
niralilocal@pragationline.com

Nirali Prakashan : S. No. 28/27, Dhyari, Near Pari Company, Pune 411041
Tel : (020) 24690204 Fax : (020) 24690316
Email : dhyari@pragationline.com,
bookorder@pragationline.com

MUMBAI

Nirali Prakashan : 385, S.V.P. Road, Rasdhara Co-op. Hsg. Society Ltd.,
Girgaum, Mumbai 400004, Maharashtra
Tel : (022) 2385 6339 / 2386 9976, Fax : (022) 2386 9976
Email : niralimumbai@pragationline.com

➤ DISTRIBUTION BRANCHES

JALGAON

Nirali Prakashan : 34, V. V. Golani Market, Navi Peth, Jalgaon 425001,
Maharashtra, Tel : (0257) 222 0395,
Mob : 94234 91860

KOLHAPUR

Nirali Prakashan : New Mahadvar Road, Kedar Plaza, 1st Floor Opp. IDBI Bank
Kolhapur 416 012, Maharashtra. Mob : 9850046155

NAGPUR

Nirali Prakashan : Above Maratha Mandir, Shop No. 3, First Floor,
Rani Jhanshi Square, Sitabuldi,
Nagpur 440012, Maharashtra Tel : (0712) 254 7129

DELHI

Nirali Prakashan : 4593/15, Basement, Aggarwal Lane, Ansari Road,
Daryaganj, Near Times of India Building, New Delhi 110002
Mob : 08505972553

BANGALURU

Nirali Prakashan : Maitri Ground Floor, Jaya Apartments, No. 99, 6th Cross,
6th Main, Malleswaram, Bangaluru 560 003, Karnataka
Mob : +91 9449043034
Email: niralibangalore@pragationline.com

www.pragationline.com　　　　　　　　　　　　　**info@pragationline.com**

PREFACE

We are very much delighted to present this book '**Taxonomy of Angiosperms'** for B.Sc. Part-I students as per new revised CBCS pattern syllabus of Punyashlok Ahilyadevi Holkar Solapur University, Solapur, June 2019.

The book covers all aspect of syllabus of Botany, B.Sc. part-I, Semester-II. We have tried to present the topics in simple words. The illustrative diagrams are given wherever necessary. At the end of each unit Questions are given for practice purpose.

The information of Botany Paper-IV is compiled in such a way that it meets all the demands of students. We have referred standard books and articles to prepare this book. We are grateful to authors and publishers of these reference books.

Authors are grateful to Management and Principals of Shankarrao Mohite Mahavidyalaya, Akluj; Santosh Bhimrao Patil College of Arts, Commerce & Science, Mandrup; and Karmvir Bhaurao Patil College, Pandharpur for their continuous encouragement and guidance.

We are thankful to the faculty members of all the colleges of Solapur University for their constant support.

We are thankful to Shri. Dineshbhai Furia, Shri. Jignesh Furia and the entire staff of **Nirali Prakashan** for taking keen interest in publishing this book and bringing out in attractive form and well in time.

The authors welcome suggestions for improvement from the readers.

– **Authors**

Nature of Question Paper

Time : 2 hrs. 30 min.   **Botany Paper – IV**   **Max. Marks = 40**

Instructions:
- i.   *All questions are compulsory.*
- ii. *Draw neat labeled **diagram** and give **captions** wherever necessary.*
- iii. *Figures to the **right** indicate **full marks**.*
- iv. *Use of logarithmic table & calculator is allowed.*

(Atomic weights: B = 1; C = 12; O = 16; N = 14; Na = 23; Cl = 35.5)

Q.1: Rewrite the following sentence by choosing the correct alternative.

	08 Marks
1.	01
2.	01
3.	01
4.	01
5.	01
6.	01
7.	01
8.	01

Q.2: Answer **any four** of the following.   **08 Marks**

1.	02
2.	02
3.	02
4.	02
5.	02
6.	02

Q.3: **A)** Write notes on **any one** of the following.   **03 Marks**

1.	03
2.	03

B) Solve/Short answer (compulsory).   **05 Marks**

Q.4: Answer **any two** of the following.   **08 Marks**

1.	04
2.	04
3.	04

Q.5: Answer **any one** of the following.   **08 Marks**

1.	08
2.	08

❖ ❖ ❖

SYLLABUS

CONTENTS

Unit 1

INTRODUCTION

Contents

1.1 INTRODUCTION

Taxonomy (Greek, taxis= arrangement, nomous= law, rule) means the **"arrangement by rules"** or "lawful arrangement". The botanists agreed on a nomenclature- arrangement of plants into convenient groups for the proper and easy handling of vast number of plants in a suitable process following certain principles or rules.

The term Taxonomy was first introduced in plant science by A. P. de Candolle (1813), a French botanist, as the theory of plant classification. Later on and till date, it is considered as a part of plant science which includes identification, nomenclature and classification.

According to G.H.M. Lawrence (1955) **"Taxonomy is a science which includes identification, nomenclature and classification of objects and is usually restricted to objects of biological origin"**. When the taxonomy is concerned with plants, it is referred to as systematic botany.

In respect to plant, both the terms are considered as synonym but are not accepted by all. Both the terms are often used variously, like chemotaxonomy (based on chemical content), cytotaxonomy (based on chromosome structure and numbers), biosystematics (systematics of living organisms), etc.

Initially, the taxonomy was based on a few macro morphological information's like habit, sex organ, etc. i.e., the artificial system. Later on, the classification was developed after considering many morphological characteristics, the natural systems. During the post-Darwinian period the taxonomy was based on evolutionary relation-ships i.e., the phylogenetic systems.

However, modern taxonomy is not restricted on morphology only. It depends on other branches of botany for good information like anatomy, cytology, physiology, phytochemistry, genetics, embryology, ecology etc.

The term 'taxon' was first introduced by Adolf Meyer (1926), a German biologist, for the animal groups. Later, in 1948, it was Herman J. Lam who proposed the term in plant science and it was accepted in the Seventh International Congress (1950). The term "taxon" indicates a taxonomic group like a variety, species, genus or any higher group.

1.1.1 Taxonomy and Systematics

There is no vocal opinion in the meaning and use of both the terms taxonomy and systematics. Mason (1950) considered taxonomy as a vast field in biological science including four main streams. These are systematics i.e., comparative study of the organisms, taxonomic systems, nomenclature, and documentation.

On the other hand, authors like Simpson (1961), Heywood (1967), Ross (1974) and many others have treated systematics as a field which covers the study of diversity, differentiation and the relationship that exists among the organisms. According to them, taxonomy is a part of systematics.

According to Solsbrig (1966), taxonomy includes nomenclature and classification but tilt massively on systematics for its concept. Later, Small (1989) reviewed the above and defined systematics as 'The science of organization and pattern of heritable relationships among the kinds and diversity of organisms' and, on the other hand, taxonomy as 'a very substantial but imprecisely separated part of systematics, that is especially concerned with the production of formal classifications of living things on the basis of genetic relationships'.

Finally, it can be concluded that due to loose and interchangeable use in past, a proper explanation at present is very difficult. The literal meaning of Taxonomy in Greek is putting in order or lawful arrangement and Systematics means putting together. Lam (1959) and Turrill (1964) boldly expressed their opinion to treat these terms synonymously and later it was followed by many others.

1.2 AIMS AND PRINICPLES OF TAXONOMY

The aim of taxonomy includes three aspects 1. Identification, 2. Nomenclature, and 3. Classification of plants. But to develop the construction of the above, the taxonomists approach in different ways.

These are of two types:

1.　Empirical approach, and;

2.　Interpretative approach.

1.　Empirical approach: This type of approach is based on the thorough observation and characterization of the organism which finally leads to the construction of a classification. Due to thorough observation of many characteristics of the specimens and development of a classification, this type becomes popularly acceptable. This type of approach was developed during pre-Darwinian period.

2.　Interpretative approach: This type of approach of classification is based on the interpretation of evolution of a taxon. This is called phylogenetic classification, which is developed during post-Darwinian period (i.e., after the publication of 'The Origin of Species by Means of Natural Selection' [1859]). This type of classification needs the data from the past history of a taxon.

The modern taxonomy made an attempt to merge the above two approaches.

The aims of modern taxonomy are:

1. To supply an appropriate method of identification.

2. To contribute classification based on natural affinities of the specimens.

3. To contribute a catalogue of taxa by studying the different flora.

4. To trace the evolution after proper observation and interpretation.

5. To supply an integrating and synthetic role to maintain the relationship among the different biological fields.

6. To provide a convenient method of identification and communication. A workable classification having the taxa arranged in hierarchy, detailed and diagnostic descriptions are essential for identification.

7. To provide an inventory of the world's flora. Although a single 'world flora' is difficult to come by, floristic records of continents (Continental Floras; *cf. Flora Europaea* by Tutin et al.), regions or countries (Regional Floras; *cf. Flora of British India* by J. D. Hooker) and states or even countries (Local Floras; *cf. Flora of Solapur* by S. P. Gaikwad & K. U. Garad) are well documented.

8. To detect evolution at work; to reconstruct the evolutionary history of the plant kingdom, determining the sequence of evolutionary change and character modification.

9. To provide a system of classification which depicts the evolution within the group. The phylogenetic relationship between the groups is commonly depicted with the help of a phylogram, wherein the longest branches represent more advanced groups and the shorter, nearer the base, primitive ones.

10. To provide an integration of all available information. To gather information from all the fields of study, analyzing this information

using statistical procedures with the help of computers, providing a synthesis of this information and developing a classification based on overall similarity.

11. To provide an information reference, supplying the methodology for information storage, retrieval, exchange and utilization.

12. To provide significantly valuable information concerning endangered species, unique elements, genetic and ecological diversity.

13. To provide new concepts, reinterpret the old, and develop new procedures for correct determination of taxonomic affinities, in terms of phylogeny and phenetics.

14. To provide integrated databases including all species of plants (and possibly all organisms) across the globe. Several big organizations have come together to establish online searchable databases of taxon names, images, descriptions, synonyms and molecular information.

The activities of plant systematics are basic to all other biological sciences and, in turn, depend on the same for any additional information that might prove useful in constructing a classification. These activities are directed towards achieving the above mentioned aims.

1.2.1 Identification, Nomenclature, and Classification

A) Identification:

Identification or determination is recognizing an unknown specimen with an already known taxon, and assigning a correct rank and position in an extant classification. In practice, it involves finding a name for an unknown specimen. This may be achieved by visiting a herbarium and comparing unknown specimen with duly identified specimens stored in the herbarium. Alternately, the specimen may also be sent to an expert in the field who can help in the identification.

Identification can also be achieved using various types of literature such as Floras, Monographs or Manuals and making use of identification keys provided in these sources of literature. After the unknown specimen has been provisionally identified with the help of a key, the identification can be further confirmed by comparison with the detailed description of the taxon provided in the literature source.

A method that is becoming popular over the recent years involves taking a photograph of the plant and its parts, uploading this picture on the website and informing the members of appropriate electronic Lists or Newsgroups, who can see the photograph at the website and send their comments to the enquirer. Members of the organization could thus help each other in identification in a much efficient manner.

B) Nomenclature:

Nomenclature deals with the determination of a correct name for a taxon. There are different sets of rules for different groups of living organisms. Nomenclature of plants (including fungi) is governed by the International Code of Botanical Nomenclature (ICBN) through its rules and recommendations. Updated every six years or so, the Botanical Code helps in picking up a single correct name out of numerous scientific names available for a taxon, with a particular circumscription, position and rank.

To avoid inconvenient name changes for certain taxa, a list of conserved names is provided in the Code. Cultivated plants are governed by the International Code of Nomenclature for Cultivated Plants (ICNCP), slightly modified from and largely based on the Botanical Code.

C) Classification:

Classification is an arrangement of organisms into groups on the basis of similarities. The groups are, in turn, assembled into more inclusive groups, until all the organisms have been assembled into a single most inclusive group. In sequence of increasing inclusiveness, the groups are assigned to a fixed hierarchy of categories such as species, genus, family, order, class, and division, the final arrangement constituting a system of classification.   The process of classification includes assigning appropriate position and rank to a new taxon (a taxonomic group assigned to any rank; pl. taxa), dividing a taxon into smaller units, uniting two or more taxa into one, transferring its position from one group to another and altering its rank. Once established, a classification provides an important mechanism of information storage, recovery and practice.

1.2.2 Principles of Taxonomy

Taxonomy is the oldest branch of Botany and was practiced in many countries like India, Greece, Rome, China, and England from long back. During early part, taxonomy was mainly aimed to develop some convenient methods of classification. One of the earliest known Indian works dealing with plants in a scientific manner is Vrikshayurveda.

Later, it was completed by Parasar before Christ. This book contained a classification based on comparative morphology of plants. The artificial system reached its climax in Carolus Linnaeus (1707-1778), the Father of modern taxonomy. Later, Michel Adanson (1727-1806) was perhaps the first person to reject all artificial systems in favour of natural system.

He proposed the idea that all characteristics are important in grouping of plants, which in the recent years came as Numerical Taxonomy. At that time, it was very difficult to consider all the characteristics of a plant in grouping of plants. For the above problem, importance was given on flower characteristics. In this system, a group of characteristics are considered in grouping of plants.

The natural system reached its climax in George Bentham and Joseph Dalton Hooker in their book 'Genera Plantarum' in July 1862 and the last part in April 1883. Before the publication of their book, the phylogenetic concept came in focus after the publication of Darwin's concept of 'Origin of Species' (1859).

This system is still continuing with eminent personalities like Engler (1886-1892), Hutchinson (1926-1973), Takhtajan (1969, 1980), Cronquist (1968, 1981) and many others. Thus, the principles have evolved through time.

However, Cronquist (1968) has formulated certain basic principles of Taxonomy in "Evolution and Classification of Flowering Plants":

1. Taxa are properly established on the basis of multiple correlations of characters.

2. Taxonomic importance of a character is determined by how well it correlates with other characters. This means that the taxonomic importance of a character is determined by a posteriori rather than a priori.

3. An important feature of taxonomy is its predictive value.

The most important principle of taxonomy is the multiple correlations of characters. To find out correlation of taxa with others, single character should not be considered, it should always be considered along with other characters.

The selected characters should show maximum correlation with other characters. The significance of selected characters depends on the degree of correlation. Characters with no distinct correlation with other characters are often accepted as an anomalous one and usually it is not important from the taxonomic point of view.

Thus, the regular flower (*Scoparia dulcis*) of Scrophulariaceae; exstipulate leaves (*Corculum leptopus*) of Polygonaceae; zygomorphic flower (*Delphinium ajacis*) of Ranunculaceae; gamopetalous corolla (Cucurbits) of Cucurbitaceae; and many such exceptional characters are taken as anomalous, because they are not correlated with common characters of taxa.

The above examples clearly indicate that they are not significant taxonomically, but are important from their identification. But the anomalous characters often help the taxa with other groups of plants where these same characters are normally available.

EXERCISE

Q.1 Multiple choice questions: **(01 Mark)**

1. The term Taxonomy was first introduced in plant science by..............

   a) A. P. de Candolle
   b) G. H. M. Lawrence
   c) Adolf Meyer
   d) Herman J. Lam

2. "Evolution and Classification of Flowering Plants" was written by..............

   a) A. P. de Candolle
   b) G. H. M. Lawrence
   c) Arthur Cronquist
   d) Herman J. Lam

3. *Flora of Solapur* is written by..............

   a) S. R. Yadav & M. M. Sardesai
   b) S. P. Gaikwad & K. U. Garad
   c) V. N. Naik & A. S. Dhabe
   d) D. S. Pokale & P. G. Diwakar

4. ....is recognizing an unknown specimen with an already known taxon.

   a) Classification
   b) Nomenclature
   c) Identification
   d) Naming

5. An important feature of taxonomy is it's .... value.

   a) economic
   b) predictive
   c) natural
   d) classical

ANSWERS

1-a	2-c	3-b	4-c	5-b

Q.2 Answer the following:　　　　　(02 Marks)

1. Define taxonomy.
2. What is classification?
3. What is meant by Empirical approach?
4. What is meant by Interpretative approach?
5. Define nomenclature.

Q.3 Write notes on the following or answer the following:

(03/04 Marks)

1. What are the principles of taxonomy?
2. What are the aims of taxonomy?
3. Write short note on Identification.

Q.4 Solve/Short answer of the following:　　　　　(05 Marks)

1. Write short note on Empirical approach and Interpretative approach.
2. Explain in brief principles of Taxonomy.
3. Write in brief about 'Taxonomy and Systematics' in hands of different taxonomists.

Q.5 Answer the following:　　　　　(08 Marks)

1. Comment upon Identification, Nomenclature, and Classification in taxonomy.
2. Write an essay on introduction of taxnomy.
3. Comment upon aims of taxonomy.

Unit 2

CLASSIFICATION

Contents

2.1 INTRODUCTION

The early history of development of botanical science is nothing but a history of development of plant taxonomy. The herbalists and agriculturists of ancient times gathered some knowledge about plants which was passed on from generation to generation.

Theophrastus (372-287 BC), the Greek philosopher-scientist, placed this knowledge of plants on a scientific footing. In his "Enquiry into Plants" he dealt with the plants at large and attempted to arrange the plants in several groups. He is, therefore, called the **"Father of Botany"**.

Linnaeus and the botanists before him tried to classify the plant kingdom using a single or a few characters chosen arbitrarily. They only thought about the convenience of following a system of classification solely to identify a particular plant. Such systems are, therefore, called artificial systems of classification. Owing to the efforts of Linnaeus the study of Botanical science entered the modern age and Linnaeus is rightly called the **"Father of Modern Botany"**.

Later on plant-taxonomists conceived the idea that the plants belonged to some natural groups and they tried to designate and distinguish such groups and tried to classify the plant kingdom accordingly. Such systems are known as natural systems of classification.

(2.1)

Such natural grouping gives the idea that the individuals under one particular group are closely related to one another although they believed in the fixity of species and simultaneous creation of all the groups by God.

After the publication of the theory of Organic Evolution by Charles Darwin and Alfred Wallace the taxonomist began to think about the origin of each of the natural groups from a more primitive group, or from an individual of a more primitive group.

In other words, some of the natural groups are more primitive and some are more recent or advanced and the recent groups have been derived from some comparatively primitive group. A system of classification based on the idea of organic evolution attempting to find out the relation between the different groups, i.e., to trace the phylogeny of the groups, is called the phylogenetic system of classification.

"**Classification** is the scientific categorization of the organisms in hierarchical series of groups. The species is generally considered as smallest group. More similar species are grouped together into a genus, similar genera grouped into family, families into an order, orders into class, similar classes into a division, and divisions into a kingdom".

2.2 TYPES OF CLASSIFICATION: ARTIFICIAL, NATURAL AND PHYLOGENETIC

Classification is an arrangement of organisms into groups on the basis of similarities. The groups are, in turn, assembled into more inclusive groups, until all the organisms have been assembled into a single most inclusive group. In sequence of increasing inclusiveness, the groups are assigned to a fixed hierarchy of categories such as species, genus, family, order, class, and division, the final arrangement constituting a system of classification. The process of classification includes assigning appropriate position and rank to a new taxon (a taxonomic group assigned to any rank; pl. taxa), dividing a taxon into smaller units, uniting two or more taxa into one, transferring its position from one group to another and altering its rank. Once established, a classification provides an important mechanism of information storage, recovery and practice.

1. Artificial Classification:

The earliest systems of classification which remained dominant from 300 B.C. up to about 1830 were artificial systems, which were based on one or a few easily observable characters of plants, such as habit (trees, shrubs, herbs, etc.) or floral characters (particularly the number of stamens and carpels).

Such type of classification using some arbitrary or at least easily observable characters, often irrespective of their affinity, is called artificial.

The sexual system of Linnaeus is a good example of artificial classification, which uses only one attribute i.e. the number of stamens for grouping plants into 24 Classes as a result of which, various unrelated taxa, which are not at all related but, similar in one respect only, have been placed under the same Class.

2. Natural Classification:

These systems of classifications are based upon overall resemblances, mostly in gross morphology, thus, utilizing as many taxonomic characters as possible, to group taxa.

Charles Darwin's proposed theory of evolution (1859) postulates that, the present day plants have descended from those existing in the ancient past, through a series of modifications in response to changing environmental conditions, which means that all present day plants are related to each other in one way or another.

Thus, the closely related plants should naturally be grouped together. This is called natural classification. Thus, larger the number of characters shared by different taxa, the more closely related they are to each other. This is the basis of modern classification.

3. Phylogenetic Classification:

The classification systems proposed after Darwin's theory is mostly phylogenetic i.e. they use as many taxonomic characters as possible in addition to the phylogenetic (evolutionary) interpretations. These are expressed in the form of phylogenetic trees or shrubs showing presumed evolution of the groups.

The natural systems are two-dimensional i.e. based on the data available at any time and is known as Horizontal Classification, whereas the addition of the third dimension i.e. past history or ancestral history results in phylogenetic classification also known as Vertical Classification or Evolutionary Classification.

2.3 BENTHAM AND HOOKER'S SYSTEM OF CLASSIFICATION

The system of classification of seed plants presented by Bentham and Hooker, two English botanists, represented the most well developed natural system. The classification was published in a three-volume work *Genera Plantarum* (1862-83).

George Bentham (1800-1884) was a self-trained botanist. He was extremely accomplished and wrote many important monographs on families such as Labiatae, Ericaceae, Scrophulariaceae and Polygonaceae. He also published Handbook of British Flora (1858) and Flora *Australiensis* in 7 volumes (1863-78).

Sir J. D. Hooker (1817-1911), who succeeded his father William Hooker as Director, Royal Botanic Gardens in Kew, England was a very well known botanist, having explored many parts of the world. He published *Flora of British India* in 7 volumes (1872-97), *Student's Flora of the British Islands* (1870).

Outline of the system of classification presented by Bentham and Hooker in *Genera plantarum* (1862-1883):

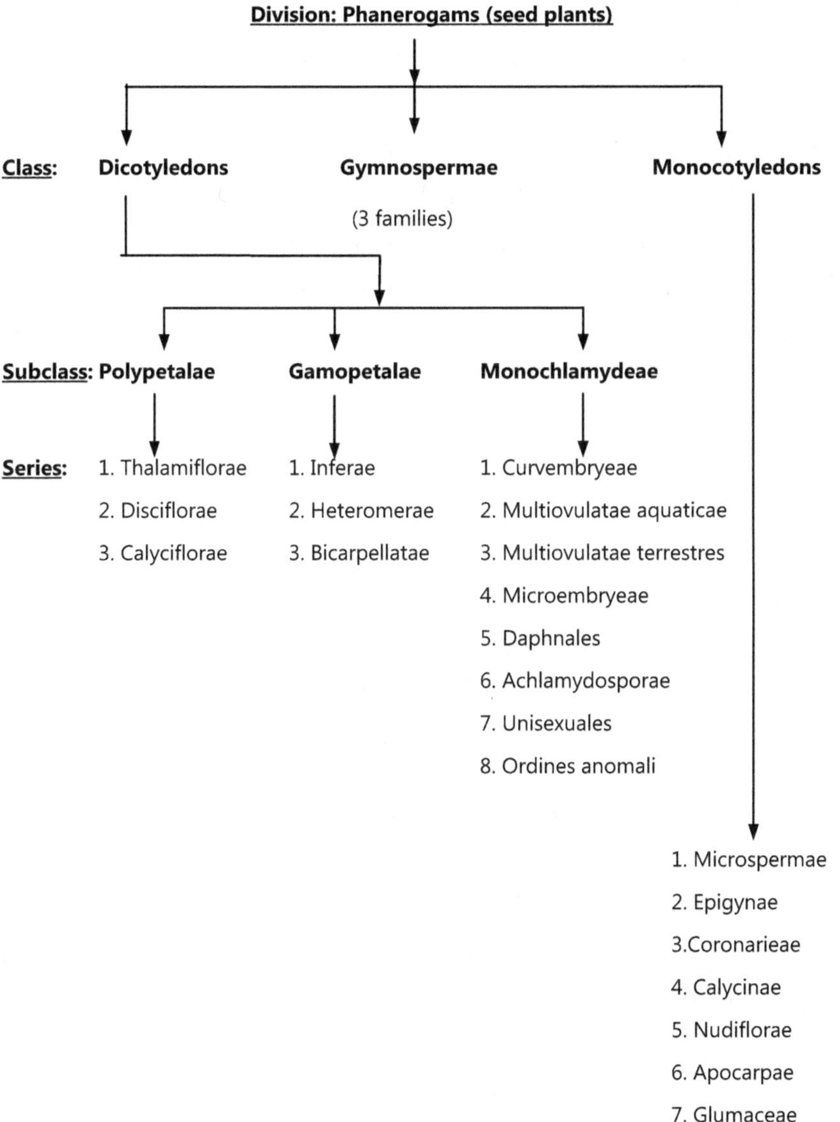

Salient features of Bentham and Hooker's system:

1. It is a classification of only the "seed plants" or Phanerogams.

2. They described about 97,205 species of seed plants belonging to about 7,569 genera of 202 families starting from Ranunculaceae up to Gramineae (Poaceae).

3. They classified all the seed plants into 3 groups or classes i.e. Dicotyledons (165 families), gymnosperms (3 families) and monocotyledons (34 families).

4. They included doubtful orders among 'Ordines Anomali' which they could not place satisfactorily.

5. Monocotyledons were described after the Dicotyledons.

6. The Dicotyledons were divided into 3 sub-classes (Polypetalae, Gamopetalae and Monochlamydeae) and 14 series. Each series again divided into cohorts (modern orders) and cohorts into orders (modern families).

7. The authors did not mention anything about the origin of the angiosperms.

8. Creation of the Disciflorae, a taxon not described by the earlier taxonomists.

9. Among the Monochlamydeae, major taxa, like the series, were divided on the basis of terrestrial and aquatic habits.

10. Polypetalae carries 82 families, 2610 genera & 31,874 species. Gamopetalae carries 45 families 2619 genera & 34,556 species. Monochlamydeae includes 36 families, 801 genera & 11,784 species. Similarly Monocotyledons consists 34 families, 1495 genera and 18,576 species.

2.4 MERITS AND DEMERITS OF BENTHAM AND HOOKER'S SYSTEM

A. Merits:

1. Each plant has been described either from the actual specimen or preserved herbarium sheets so that the descriptions are detailed as well as quite accurate.

2. The system is highly practical and is useful to students of systematic botany for easy identification of species.

3. The flora describes geographical distribution of species and genera.

4. The generic descriptions are complete, accurate and based on direct observations.

5. Larger genera have been divided into sub genera, each with specific number of species.

6. Dicots begin with the order Ranales which are now universally considered as to be the most primitive angiosperms.

7. Placing of monocots after the dicot is again a natural one and according to evolutionary trends.

8. The placing of series disciflorae in between thalami florae and calyciflorae is quite natural.

9. The placing of gamopetalae after polypetalae is justified since union of petals is considered to be an advanced feature over the free condition.

B. Demerits:

1. Keeping Gymnosperms in between dicots and monocots is anomalous.

2. Subclass monochlamydeae is quite artificial.

3. Placing of monochlamydeae after gamopetalae does not seem to be natural.

4. Some of the closely related species are placed distantly while distant species are placed close to each other.

5. Certain families of monochlamydeae are closely related to families in polypetalae, e.g. Chenopodiaceae and Caryophyllaceae.

6. Advanced families, such as Orchiadaceae have been considered primitive in this system by placing them in the beginning. Placing of Orchidaceae in the beginning of monocotyledons is unnatural as it is one of the most advanced families of monocots. Similarly, Compositae (Asteraceae) has been placed near the beginning of gamopetalae which is quite unnatural.

7. Liliaceae and Amaryllidaceae were kept apart merely on the basis of characters of ovary though they are very closely related.

8. There were no phylogenetic considerations.

EXERCISE

Q.1　Multiple choice questions:　　　　　　　　**(01 Mark)**

1. .... called the "Father of Botany".

   a) Linnaeus　　　　　　　b) Charles Darwin

   c) Bentham　　　　　　　d) Theophrastus

2. .... is rightly called the "Father of Modern Botany".

   a) Linnaeus　　　　　　　b) Charles Darwin

   c) Bentham　　　　　　　d) Theophrastus

3. Classification using some arbitrary or atleast easily observable characters, often irrespective of their affinity, is called..............

   a) natural　　　　　　　b) artificial

   c) phylogenetic　　　　　d) modern

4. The order .... which are now universally considered as to be the most primitive angiosperms.

   a) Ranales　　　　　　　b) Rosales

   d) Myrtales　　　　　　　d) Unisexuales

5. System of classification of seed plants presented by Bentham and Hooker was published in..............

   a) Enquiry into Plants　　　b) Handbook of British Flora

   c) Genera Plantarum　　　d) Flora of British India

ANSWERS

1-d	2-a	3-b	4-a	5-c

Q.2 Answer the following:　　　　　　　　**(02 Marks)**

1. Define natural classification.

2. Define artifical classification.

3. Define phylogenetic classification.

4. Enlist the series of sub-class Ploypetalae.

5. Enlist the series of sub-class Gamopetalae.

Q.3 Write notes on the following or answer the following:

(03/04 Marks)

1. Write a note on artifical cassification.

2. Write a note on natural cassification.

3. Write a note on phylogenetic cassification.

Q.4 Solve/Short answer of the following: (05 Marks)

1. Write down the salient features of Bentham and Hooker's system of classification.

2. Write down the merits of Bentham and Hooker's System of classification.

3. Write down the demerits of Bentham and Hooker's System of classification.

Q.5 Answer the following: (08 Marks)

1. Explain the types of classification.

2. Comment upon outline of the system of classification presented by Bentham and Hooker.

3. Comment upon merits and demerits of Bentham and Hooker's system of classification.

Unit **3**

IDENTIFICATION AND NOMENCLATURE

Contents

3.1 IDENTIFICATION OF PLANTS

Identification or determination is recognizing an unknown specimen with an already known taxon, and assigning a correct rank and position in an extant classification. In practice, it involves finding a name for an unknown specimen. This may be achieved by visiting a herbarium and comparing unknown specimen with duly identified specimens stored in the herbarium. Alternately, the specimen may also be sent to an expert in the field who can help in the identification. Identification can also be achieved using various types of literature such as Floras, Monographs or Manuals and making use of identification keys provided in these sources of literature.

After the unknown specimen has been provisionally identified with the help of a key, the identification can be further confirmed by comparison with the detailed description of the taxon provided in the literature source. A method that is becoming popular over the recent years involves taking a photograph of the plant and its parts, uploading this picture on the website and informing the members of appropriate electronic Lists or Newsgroups, who can see the photograph at the website and send their comments to the enquirer. Members of the fraternity could thus help each other in identification in a much efficient manner.

Characters considered before plant identification:

1. Whether a plant is herbaceous or woody, and annual or perennial in nature.

2. Whether or not milky or colored sap is present in the leaf, stem or other plant parts.

3. The leaf type, phyllotaxy, and venation.

4. Presence or absence and type of stipule on young shoots.

5. The distribution and kinds of surface coverings (i.e. hairs, trichomes, spines, etc.).

6. The parts of the flower and the number of sepals and petals, whether separate or fused, and also their arrangement i.e. aestivation

7. Whether perianth is present in one or more series, or absent.

8. Whether pappus (e.g. Asteraceae) or epicalyx (e.g. Malvaceae) or similar structures are present.

9. Whether a nectar-secreting disc is present in the flowers (e.g. Rutaceae).

10. Whether the flowers are actinomorphic or zygomorphic.

11. The number and attachment of stamens and if there is any fusion of anthers or filaments.

12. The number of pistils, styles and stigmas of the gynoecium, observation of a transverse section of the ovary, the number of locules, number of ovules per locule, and also the placentation.

13. The position of the ovary and fusion of the perianth by observing a longitudinally cut section of the entire flower through its centre.

Plant Identification Methods:

The methods of identification include the following steps:-

(a) Expert Determination: The best method of identification is expert determination in terms of reliability or accuracy. In general the experts have prepared treatments (monographs, revisions, synopses) of the group in question, and it is probable that the more recent floras or manuals include the expert's concepts of taxa.

Experts are typically found in botanical gardens, herbaria, museums, colleges, universities, etc. However, although of great reliability, this method presents problems of requiring the valuable time of experts and creating delays for identification.

(b) Recognition: It approaches expert determination in reliability. This is based on extensive, past experience of the identifier with the plant group in question. In some groups this is virtually impossible.

(c) Comparison: A third method is by comparison of an unknown with named specimens, photographs, illustrations or descriptions. Although this is a reliable method, it may be very time consuming or virtually impossible due to the lack of suitable materials for comparison. The reliability is, of course, dependent on the accuracy and authenticity of the specimens, illustrations, or descriptions used in the comparison.

(d) The Use of Keys and Similar Devices (Synopses, Outlines, etc.): This is by far the most widely used method and does not require the time, materials, or experience involved in comparison and recognition.

3.2 NOMENCLATURE, BINOMIAL NOMENCLATURE OF PLANTS

Nomenclature deals with the determination of a correct name for a taxon. There are different sets of rules for different groups of living organisms. Nomenclature of plants (including fungi) is governed by the International Code of Botanical Nomenclature (ICBN) through its rules and recommendations. Updated every six years or so, the Botanical Code helps in picking up a single correct name out of numerous scientific names available for a taxon, with a particular circumscription, position and rank. To avoid inconvenient name changes for certain taxa, a list of conserved names is provided in the

Code. Cultivated plants are governed by the International Code of Nomenclature for Cultivated Plants (ICNCP), slightly modified from and largely based on the Botanical Code.

Nomenclature deals with the application of a correct name to a plant or a taxonomic group. In practice, nomenclature is often combined with identification, since while identifying an unknown plant specimen; the author chooses and applies the correct name. The favorite temperate plant is correctly identified whether you call it 'Safarchand' (vernacular Marathi name), Apple- *Pyrus malus*; but only by using the correct scientific name *Malus domestica* does one combine identification with nomenclature. The current activity of botanical nomenclature is governed by the International Code of Botanical Nomenclature (ICBN) published by the International Association of Plant Taxonomy (IAPT). The Code is revised after changes at each International Botanical Congress. Naming of cultivated plants is governed by the International Code of Nomenclature for Cultivated Plants (ICNCP), which is largely based on ICBN with a few additional provisions.

Need for Scientific Names:

Scientific names formulated in Latin are preferred over vernacular or common names since the latter pose a number of problems:

1. Vernacular names are not available for all the species known to man.

2. Vernacular names are restricted in their usage and are applicable in a single or a few languages only. They are not universal in their application.

3. Common names usually do not provide information indicating family or generic relationship. Roses belong to the genus *Rosa*; woodrose is a member of the genus Ipomoea and primrose belongs to the genus *Primula*. The three genera, in turn, belong to three different families- Rosaceae, Convolvulaceae and Primulaceae, respectively. Oak is similarly common name for the species of genus *Quercus*, but Tanbark oak is *Lithocarpus*, poison

oak a *Rhus*, silver oak a *Grevillea* and Jerusalem oak a *Chenopodium*.

4. Frequently, especially in widely distributed plants, many common names may exist for the same species in the same language in the same or different localities. Cornflower, bluebottle, bachelor's button and ragged robin all refer to the same species *Centaurea cyanus*.

5. Often, two or more unrelated species are known by the same common name. Bachelor's button, may thus be *Tanacetum vulgare, Knautia arvensis* or *Centaurea cyanus*. Cockscomb, is similarly, a common name for Celosia cristata but is also applied to a seaweed *Ploca-mium coccinium* or to *Rhinanthus minor*.

Therefore, to overcome this indistinctness problem happens due to common, vernacular, local or English name of plant species, there is need to only one correct 'scientific name' for each plant species or rank.

Why Latin?

Scientific names are treated as Latin regardless of their origin. It is also mandatory to have a Latin diagnosis for any new taxon published 1st January 1935 onwards. The custom of Latinized names and texts originates from medieval scholarship and custom continued in most botanical publications until the middle of nineteenth century. Descriptions of plants are not written in classical Latin of Cicero or of Horace (Sense-for-sense translation/word-for-word translation), but in the 'lingua-franca' (i.e. bridge language) spoken and written by scholars during middle ages, based on popular Latin spoken by ordinary people in the classical times. The selection has several advantages greater than modern languages:

(i) Latin is a dead language and as such meanings and interpretation are not subject to changes unlike, English and other languages;

(ii) Latin is specific and exact in meaning;

(iii) Grammatical sense of the word is commonly obvious (white translated as album-neuter, alba-feminine or albus- masculine); and

(iv) Latin language employs the Roman alphabet, which fits well in the text of most languages.

Binomial Nomenclature:

The binomial nomenclature is the system where naming of plants consists of two words- a generic name and a specific epithet (name). The first one is the generic name and the second one is the specific epithet (name). Both the names together form a binary or binomial name.

Linnaeus for the first time proposed that every living being has binomial name, i.e., a name with two epithets. One is generic and the other is specific epithet. If an organism has a variety also, then the name becomes trinomial. Linnaeus proposed some rules for generic names of plants in Fundamental Botanica (1736) and Critica Botanica (1737).

A.P.de Candolle for the first time proposed rules for nomenclature of plants which are passed by International Botanical Congress at Paris (1867). For the first time it was a Swedish Naturalist Carolus Linnaeus who started naming plants in 1753 as Binomial names. It was published in his book "Species plantarum".

In a binomial, only the generic name should start with capital letter and all others in small letters. After selecting the name of a particular plant, it must be added with the name of the author. If the author's name is too long, it should be mentioned in abbreviated form. The names of the plants are written in Latin. The scientific name i.e., the Latin name of 'Bhimapushapa' is *Crinum solapurense*. The first name, *Crinum* is the generic name and the second name *solapurense* is the specific epithet (name).

To complete the name, the author's name in abbreviated form should be added at the end. So the complete scientific name of

'Bhimapushapa' is *Crinum solapurense* Gaikwad et. al The Gaikwad et.al indicates the name of Gaikwad & his co-worker i.e. S. P. Gaikwad, K. U. Garad & R. D. Gore, who has given the name.

Binomials should be typed in Italic type; or in case of handwriting, both generic and specific epithet should be underlined separately.

3.3 PRINCIPLES OF ICBN

ICBN refers to the "International Code of Botanical Nomenclature". Name is the means of reference to all living and non-living things. Any object known to human being is given a name to describe and communicate ideas about it. The name may be different in different languages and at different places. The art of naming the object is known as Nomenclature. And when it comes to naming of plants it is called Botanical nomenclature.

Introduction of ICBN 1983:

The process of naming plants based on international rules proposed by botanists to ensure a stable and universal uniform system is called Botanical nomenclature.

1. Botany requires a precise and simple system of nomenclature used by Botanists in all countries, dealing, on the one hand, with the terms which denote the ranks of taxonomic groups or units, and on the other hand with the scientific names which are applied to the individual taxonomic groups of plants.

2. The purpose of giving a name to a taxonomic group is not to indicate its character or history, but to supply a means of referring it and to indicate its taxonomic rank. The code aims at the provision of a stable method of naming taxonomic groups, avoiding and rejecting the use of names which may cause error or ambiguity or throw science into confusion. It avoids the useless creation of names.

3. The Principles form the basis of the system of Botanical Nomenclature.

4. The detailed provisions are divided into Rules and Recommendations. Examples are added to the rules and the recommendations to illustrate them.

5. The object of the Rules is to put the nomenclature of the past into order and to provide for that of the future, names contrary to a rule cannot be maintained.

6. The Recommendations deal with subsidiary points, their object being to bring about greater uniformity and clearness, especially in future nomenclature, names contrary to a recommendation cannot, on that account, be rejected, but they are not examples to be followed.

7. The provisions regulating the modification of this code from its last decisions.

8. The Rules and Recommendations apply to all organisms treated as plants (except Bacteria), whether fossil or non-fossil. Special provisions are needed for certain groups of plants. The International Code of Nomenclature of cultivated plants (1980) was adopted by the International Commission for the Nomenclature of Cultivated Plants; provisions for the names of hybrids are also appearing in Appendix I.

9. The only proper reason for changing a name is either a more profound knowledge of the facts resulting from adequate taxonomic study or the necessity of giving up nomenclature that is contrary to the rules.

10. In the absence of a relevant rule or where the consequences of rules are doubtful, established custom is followed.

11. This edition of the code supersedes all previous editions.

ICBN Principles:

The International Code of Botanical Nomenclature is based on the following set of six principles, which are the philosophical basis of the Code and provide guidelines for the taxonomists who propose amendments or deliberate on the suggestions for modification of the Code: The Principles were laid down in 1983.

1. Botanical Nomenclature is independent of Zoological Nomenclature. The Code applies equally to the names of taxonomic groups treated as plants whether or not these groups were originally so treated.

2. The application of names of taxonomic groups is determined by means of nomenclatural types.

3. Nomenclature of a taxonomic group is based upon priority of publication.

4. Each taxonomic group with a particular circumscription, position and rank can bear only one correct name, the earliest that is in accordance with the rules.

5. Scientific names of taxonomic groups are treated as Latin, regardless of derivation.

6. The rules of nomenclature are retroactive, unless expressly limited.

EXERCISE

Q.1 Multiple choice questions: **(01 Mark)**

1. International Code of Botanical Nomenclature is based on the set of .... principles.

    a) five   b) six

    c) seven   d) 10

2. ....for the first time proposed that every living being has binomial name.

    a) Bentham   b) Hooker

    c) Linnaeus   d) Mayer

3. Scientific names are treated as .... regardless of their origin.

   a) Latin   b) English

   c) Hindi   d) Marathi

4. The binomial nomenclature is the system where naming of plants consists of .... words.

   a) three   b) five

   c) four   d) two

5. The best method of identification is .... in terms of reliability or accuracy.

   a) Recognition   b) Expert Determination

   c) The Use of Keys   d) Comparison

ANSWERS

1-b	2-c	3-a	4-d	5-b

Q.2 Answer the following: (02 Marks)

1. Define identification.

2. What is binomial nomenclature?

3. What is Expert Determination?

4. What is Recognition?

5. What is Comparison?

Q.3 Write notes on the following or answer the following:

(03/04 Marks)

1. Write short note on Binomial nomenclature of plants.

2. Why scientific names are treated as Latin?

3.   Write short note Need for Scientific Names.

Q.4 Solve/Short answer of the following: (05 Marks)

1. Why scientific names are treated as Latin?

2. Explain in brief need for 'scientific names' to plant species.

3. Which are the methods used for plant identification?

Q.5 Answer the following: (08 Marks)

1. Comment up on characters considered before plant identification.

2. Write an essay on introduction and principles of ICBN.

3. Write an essay on Identification of plants.

Unit 4

HERBARIUM AND BOTANICAL GARDEN

Contents

4.1 HERBARIUM - STEPS IN PREPARATION AND SIGNIFICANCE

4.1.1 Herbarium

Herbarium (plural- Herbaria) is defined as a place where the collection of plants, that usually have been dried, pressed, preserved on sheets and arranged according to any accepted system of classification for future reference and study.

In fact, it is a great clean system for information about plants, both primary in the form of actual specimens of the plants, and secondary in the form of published information, pictures and recorded notes.

Method for (steps in) preparation of herbarium specimens:

The preparation of a herbarium involves following steps:

1. Field visits (work)

2. Collection of specimens

3. Pressing and Drying

4. Mounting on a herbarium sheet

5. Poisoning

6. Labeling

7. Proper storage

1. Field Visit (Work) and Collection of Specimen:

The fieldwork of specimen preparation involves plant collection, pressing and partial drying of the specimens. The plants are collected for various purposes: building new herbaria or enriching older ones, compilation of floras, material for museums and class work, ethnobotanical studies, and introduction of plants in gardens. Depending on the purpose, resources, proximity of the area and duration of studies, fieldwork may be undertaken in different ways:

Collection trip: Such a trip is of short duration, usually one or two days, to a nearby place, for brief training in fieldwork, vegetation study and plant collection by groups of students.

Exploration: This includes repeated visits to an area in different seasons, for a period of a few years, for intensive collection and study, aimed at compilation of floristic accounts.

Expedition: Such a visit is undertaken to remote and difficult area, to study the flora and fauna, and usually takes several months.

The specimen collected should be as complete as possible. Herbs, very small shrubs, as far as possible, should be collected complete, in flowering condition, along with leaves and roots. Trees and shrubs should be collected with both vegetative and flowering shoots, to enable the representation of both leaves and flowers. A complete specimen possesses all parts including root system, flowers and fruits. Therefore, regular field visits are necessary to obtain information at every stage of growth and reproduction of a plant species.

In the fields, the tools required are mainly trowel (digger) for digging roots, scissors and knife for cutting twigs, a stick with a hook for collection of parts of tall trees, a field note book, polythene bag, old newspaper and magazines.

To avoid damage during transportation and preservation at least 5 specimens (twigs) of a plant should by collected. The collected specimens (twigs) are transported in a vasculum (specimen box) to prevent willing, livery collected specimen must be tagged with a field number and necessary information should be recorded in a field note book.

2. Pressing and Drying:

The specimens are spread out between the folds of old newspapers or blotting sheets avoiding overlapping of parts. The larger specimen may folded in 'N' or' W' shapes. The blotting sheets with plant specimen should be placed in the plant press for drying. After 24 to 48 hrs the press is opened, depending on season.

3. Mounting on a herbarium sheet:

The dried specimens are mounted on herbarium sheets of standard size (41 × 29 cm). Mounting is done with the help of glue, adhesive or cello-tape etc. The bulky plant parts like dry fruits seeds, cones etc. are dried without pressing and are put in small envelops called fragment packets. Succulent plants are not mounted on herbarium sheets but are collected in 4% formalin or FAA (Formalin Acetic Alcohol).

4. Poisoning:

The mounted specimens are sprayed with fungicides like 2% solution of mercuric chloride ($HgCl_2$).

5. Labeling:

A label is pasted or printed on the lower side at right hand corner of the herbarium sheet. The label should indicate the information about the complete scientific name, family, common/local name, locality, altitude (GPS), habit & habitat, date of collection, name of collector, etc.

<table>
<tr><td colspan="2" align="center">DEPARTMENT OF BOTANY

SANTOSH BHIMRAO PATIL ARTS, COMMERCE & SCIENCE COLLEGE, MANDRUP

 </td></tr>
<tr><td>Date:</td><td>Collection No:</td></tr>
<tr><td colspan="2">Botanical name:</td></tr>
<tr><td colspan="2">Family:</td></tr>
<tr><td colspan="2">Common/Local name:</td></tr>
<tr><td colspan="2">Locality:</td></tr>
<tr><td colspan="2">GPS:</td></tr>
<tr><td colspan="2">Habit & Habitat:</td></tr>
<tr><td colspan="2">Field note:</td></tr>
<tr><td colspan="2">Collected by:</td></tr>
<tr><td colspan="2">Identified by:</td></tr>
</table>

Fig. 4.1: Format of Label

6. Storage:

Properly dried, pressed and identified plant specimens are placed in thin paper folders (specimen covers; usually white color) which are kept together in thicker paper folders (genus covers; usually faint purple color), and finally they are incorporated into the herbarium cupboards in their proper position according to a well known system of classification.

In India, herbarium specimens in herbarium cupboards are arranged according to the Bentham and Hooker's system of classification habitually. Type specimens are generally stored in separate and safe places.

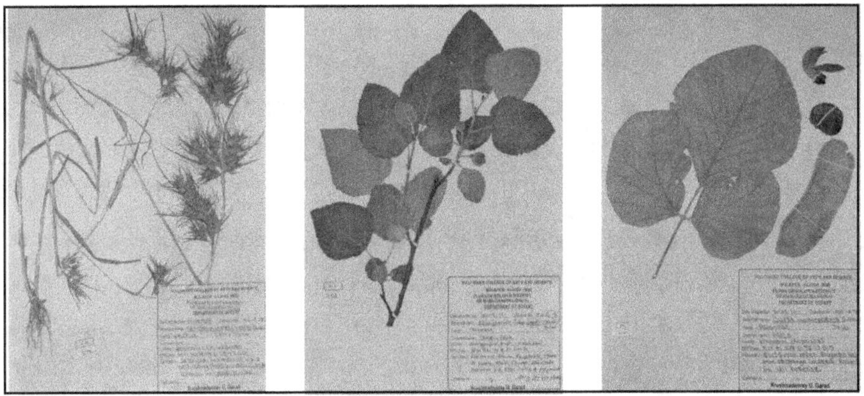

Fig. 4.2: A sample herbarium sheet with mounted specimen and a label

4.1.2 Significance of Herbarium

From a safe place for storing pressed specimens, especially type material, herbaria have gone a long way in becoming major centers of taxonomic research. Additionally, herbaria also form an important link for research in other fields of study. The classification of the world flora is primarily based on herbarium material and associated literature. More recently, the herbaria have gained importance for sources of information on endangered species and are of primary interest to conservation groups. The major roles played by a herbarium include:

1. **Storehouse of plant specimens:** Primary role of a herbarium is to store dried plant specimens, safeguard these against loss and destruction by insects, and make them available for study.

2. **Safe custody of type specimens:** Type specimens are the principal proof of the existence of a species or an infraspecific taxon. These are kept in safe custody, often in rooms with restricted access, in several major herbaria.

3. **Compilation of Floras, Manuals and Monographs:** Herbarium specimens are the 'original documents' upon which the knowledge of taxonomy, evolution and plant distribution rests. Floras, manuals and monographs are largely based on herbarium resources.

4. **Training in herbarium methods:** Many herbaria carry facilities for training graduates and undergraduates in herbarium practices, organizing field trips and even expeditions to remote areas.

5. **Identification of specimens:** The majorities of herbaria have a wide-ranging collection of specimens and offer facilities for on-site identification or having the specimens sent to the herbarium identified by experts. Researchers can personally identify their collection by comparison with the duly identified herbarium specimens.

6. **Information on geographical distribution:** Major herbaria have collections from different parts of the world and, thus, scrutiny of the specimens can provide information on the geographical distribution of a taxon.

7. **Preservation of voucher specimens:** Voucher specimens preserved in various herbaria provide an index of specimens on which a chromosomal, phytochemical, ultrastructural, micro-morphological or any specialized study has been undertaken. In the case of a contradictory or doubtful report, the voucher specimens can be critically examined in order to arrive at a more satisfactory conclusion.

4.2 BOTANICAL GARDENS OF INDIA

The garden is generally defined as a place for growing flowers, fruits or vegetables. But "Botanic or Botanical garden is an educational institution for scientific workers and general public or layman to awake and enlighten interest in plant life".

The botanical gardens are of immense value not only to botanists, home gardeners, nurserymen, horticulturists, landscape gardeners and foresters but also to millions of national and international tourists.

The botanical gardens should have morphological gardens to display seed dispersal in plants; genetics or breeding garden to display the laws of heredity and a taxonomic garden to display plant families. There should be a section of economic plants; green houses and nurseries for propagating and cultivating exotic, end genetic and delicate plants.

A botanical garden is an institution for botanical research, especially on the native flora of the region. There should be a herbarium, library, photographic studies, lecture pavilion and recreational facilities. In fact all the fundamental and applied aspects of botany come within the purview of botanical garden and it becomes the centre of cultural activities of the region in which it is situated.

Functions of Botanical Gardens:

The botanical gardens are the natural source of science and culture. Functions of Botanical Gardens are as fallows.

1. Botanical gardens act as out-door laboratories.

2. Initiate studies on the tropical and temperate ecosystems and their biota, before they are lost to science and preserve such systems.

3. Serve as centers of gene pools or germplasm bank of wild relatives of economically important plants.

4. Establish Nature centers and youth Museums to focus attention on destruction of tropical and temperate ecosystem, environmental degradation.

5. Maintain less attractive and abandoned ornamental plants.

6. Train city arborists in the plantation of trees in urban areas.

7.   Collaborate the university and others to conduct research in environmental biology.

8.   Organize educational programs to create environmental awareness among children students and train teachers in environmental education.

9.   Centers of conservation for endangered and rare species.

10. Botanical gardens provide living plant materials for research.

11. They serve as pollution indicator centers by growing pollution-susceptible plants.

12. Most of the economic plants were originally introduced and distributed to the other parts of the world through botanic gardens.

13. Gardens also arrange flower shows, put on displays seasonal plants, flowers and plants of unusual interest.

14. The landscape gardens are becoming quite popular and land a great charm to the adjoining building like libraries, museums etc.

15. Conserve the flora and fauna in natural habitat.

4.2.1 Sir J. C. Bose Botanical Garden, Calcutta

The Acharya Jagadish Chandra Bose Indian Botanic Garden previously known as Indian Botanic Garden and the Calcutta Botanic Garden, is situated in Shibpur, Howrah near Kolkata. They are commonly known as the Calcutta Botanical Garden, and previously as the Royal Botanic Garden, Calcutta. The gardens exhibit a wide variety of rare plants and a total collection of over 12,000 specimens spread over 109 hectares. It is under Botanical Survey of India (BSI) of Ministry of Environment and Forests, Government of India.

Apart from being a home to 12,000 perennial plants of 1,400 species, the garden has much more to it. It has conservatories, glass houses, greenhouses and 25 divisions of herbaceous plants. The garden is the major center of horticultural and botanical research in

India. The greenery and plants are a major inspiration for educational and development research. Apart from students and researchers, the garden also seems to attract photo walkers and tourists from all over the world due to its natural opulence it offers.

Information of Garden:

1. The garden is one of the largest and oldest green reserves in South East Asia.

2. Its founding father is Lieutenant Colonel Robert Kyd and it was established in the year 1787.

3. As of now, the garden is home to over 12,000 perennial plant species and hundreds of dried plants from all over the world.

4. Apart from Robert Kyd as the first honorary superintendent, the contributions by Dr. George King, Thomas Anderson and William Roxburgh have been significant in deciding what the garden is being hailed as today.

5. The latest census was conducted in the year 2007 and according to figures revealed; there are 14,000 species of trees and 13,722 species of plants. Out of these 500 species are considered endangered or rare.

6. **"The Great Banyan Tree"** remains the prime attraction of the garden. It is the biggest banyan tree in the world and forms the second largest canopy.

7. The garden brings a rare water lily, Victoria amazonica.

8. Here, one will find trees from all over the world. The garden boasts from exotic species from Nepal, Penang, Java, Sumatra and Brazil find a place to stay. Mahogany trees, Cuban palms, mango trees, tamarind trees as well as dedicated space to orchids and cactus.

9. Coconut tree from Sicily, the mad tree, collection of aquatic plants and exotic selection of bamboo, citrus, jasmine, water lilies, ferns, creepers, Hibiscus, Ficus and Bougainvillea can be found here in their perfect state.

10. Exotic plants such as giant water lilies, bread fruit tree, double coconut, Krishna-bot-tree, the Shivalinga tree and water lilies have found a home in Kolkata Botanical Garden.

11. Serpentine Lake for boating is also open for the visitors.

4.2.2 Lead Botanical Garden of Shivaji University Kolhapur

Botanical gardens are the place of interest for local people as well as researchers. Most of the botanical gardens aim at conservation and education. They potentially offer community education about conservation, conservation attitudes, and encourage the public to support conservation efforts.

Generally society is found to be less interested in conservation issues but the places like botanic gardens, museums, zoos, aquariums, heritage sites, natural areas and wildlife tourism sites can help to motivate the activities of public. Living collection of plant is most significant asset of botanical gardens. They seem to be most ideal places for conservation of RET (Rare, Endemic, & Threatened) and economically important plants of particular region.

In this perspective Lead Botanic Garden (LBG) is wonderful concept nurtured by MoEF for conservation of RET species in the Post-CBD period under which the Botanical Garden of Shivaji University, Kolhapur (SUK) is recognized by MoEF as Lead Botanic Garden for Western India.

Objectives of Lead Botanical Garden:

1. Ex-situ conservation and multiplication of Threatened and Endemic plants species of Western Ghats.

2. Establishment of seed-banks.

3. Reintroduction and rehabilitation of plants in its natural habitats.

4. Create awareness in people about biodiversity conservation and sustainable utilization.

Functions of Lead Botanical Garden:

1. Help to conserve natural vegetation specially Threatened and Endemic species through multiplying and rehabilitating them in natural habitats.

2. Undertake botanic research resulting in excellent referral system for plants as authentically identified, classified and labeled live collection in gardens and as dry collections (pressed, processed and mounted specimens) in herbaria both for monitoring and documentation of threatened and endemic plant resources of the country.

3. Study of phenology and response of the plants to climate variability/change.

4. Carry out conservation biological studies with a view to find out ecological, biological and genetic bottlenecks or barriers in the reproduction and survival of species.
Carry out rehabilitation/recovery programs for threatened and endemic species.
Serve as center of training with expertise in a focused area of subject specialization including horticulture.

5. Building up of information on in-situ as well as ex-situ conservation of the threatened and endemic species and their habitats.

6. Compile information on the area of occurrence, area of occupancy, number and size of populations, spatial distribution of populations, identification of important associates such as pollinators and dispersers, reproductive and breeding systems, population trends in relation to habitat changes and pattern of disturbance etc. Prepare Red Data Sheets for the selected species as per IUCN format.

7. Promote environmental awareness of nature conservation through well designed education programs and educational materials.

8. Develop relevant research and development (R&D) expertise and capabilities in undertaking modern conservation and gene banking techniques including *in-vitro* tissue banks, DNA and Cryo-Bank.

EXERCISE

Q.1   Multiple choice questions: **(01 Mark)**

1. The standard size herbarium sheets are of .... cm.

   a) 40 × 29   b) 41 × 29

   c) 29 × 41   d) 30 × 50

2. Sir J. C. Bose Botanical Garden is situated in ....

   a) Calcutta   b) Goa

   c) Delhi   d) Pune

3. The mounted specimens on herbarium sheet are poisoned with the help of chemical...............

   a) NaCl   b) HCl

   c) $HgCl_2$ d) $BaCl_2$

4. Lead Botanical Garden is situated in...............

   a) Calcutta   b) Goa

   c) Kolhapur   d) Pune

ANSWERS

1-b	2-a	3-c	4-d

Q.2 Answer the following: **(02 Mark)**

1. Define Herbarium.

2. Draw format of Herbaria lebal.

3. What is Collection trip?

4. What is meant by Exploration?

5. Define poisoning of plants.

Q.3 Write notes on the following or answer the following:

(03/04 Marks)

1. Enlist the correct steps in preparation of herbarium specimens.

2. Write a short note on collection of plant specimens for herbarium.

3. Write a short note on labeling and storage of plant specimen for herbarium.

Q.4 Solve/Short answer of the following: (05 Marks)

1. Write a short note on significance of herbarium.

2. What are the functions of Lead Botanical Garden?

3. Enlist the functions of Botanical garden.

Q.5 Answer the following: (08 Marks)

1. What is Herbaria? Explain in detail method of preparation for herbarium specimens.

2. Write an essay on Sir J. C. Bose Botanical Garden, Calcutta.

3. Write an essay on Lead Botanical Garden of Shivaji University Kolhapur.

Unit 5

STUDY OF ANGIOSPERMS FAMILIES

Contents

5.1 INTRODUCTION

The idea of Plant Families is that plants which have something in common can be grouped together. Knowing which Family a plant belongs to can be useful - not just a way of showing off!

For a start, it can help identify a new plant. If your unknown plant has the characteristics of a particular Family, then you can narrow the search to find its identity.

It can give you an idea of what the plant looks like. Almost anything in the Asteraceae Family will look like a Daisy. Most members of the Campanulaceae Family have blue flowers in a bell or star shape. Many collections of seed from their natural habitat just give the Family name.

If you know which Plant Family a plant belongs to, it might help you to find the seeds. For instance, members of the Cabbage family

(Brassicaceae) have a seedpod that has a thin papery membrane between the two halves (like Honesty), members of the Nettle Family (Lamiaceae) don't have a seed pod, they have four seeds on a pad at the bottom of the open calyx, and members of the old Leguminosae Family all have their seeds in legumes (pods like pea or bean pods).

Knowing the Plant Family can tell you where the seed pod will be - on the stalk side of the flower (called an Inferior Ovary - such as in Amaryllidaceae, Cannaceae) or in the middle of the flower itself (a Superior Ovary - as in Nyctaginaceae, which includes Mirabilis, Geraniaceae, Iridaceae).

It can often tell you what the seeds will be like - whether they're large or small, and whether there are a lot of them in a seedpod or only one. Members of the Campanulaceae have many small seeds in a capsule, seeds of the Asclepiadaceae are usually flat and oval with long silky hairs, members of the Solanaceae Family have either a berry or a capsule with many seeds.

Knowing the Plant Family can also give you a clue about how to germinate any new seeds you have from other plants in that Family. I know I've had success with many members of the Geranium Family by nicking them and sowing them indoors by the Norman Deno method. That's also worked for many members of the Lily family, but many members of the Iris Family need to be sown outside and take a long time to germinate.

Knowing which Family a plant belongs to can tell you what the seedling looks like. Seeds of all the Monocot families (such as Liliaceae, Iridaceae, other bulbs, grasses and palms) will come up with only one seed leaf. Dicots (most of the other larger plant families) have two seed leaves.

5.2 SYSTEMATIC POSITION, MORPHOLOGICAL (I.E. VEGETATIVE & REPRODUCTIVE CHARACTERS) & DISTINGUISHING CHARACTERS WITH ECONOMIC IMPORTANCE OF FOLLOWING FAMILIES

5.2.1 Caesalpiniaceae

Systematic position:

Division	: Phanerogams	Flowering plants.
Sub-division	: Angiospermae	Ovules are enclosed in ovary; seeds are enclosed in fruits.
Class	: Dicotyledons	Seed with 2 cotyledons; flowers pentamerous or tetramerous; leaves with reticulate venation.
Sub-class	: Polypetalae	Sepals and petals distinct; petals free.
Series	: Calyciflorae	Flowers perigynous or epigynous.
Order	: Rosales	Usually have four or five petals and are flat or cup-shaped; fruits are fleshy.
Family	: Caesalpiniaceae	

- **Distribution:** There are about 135 genera in the family Caesalpiniaceae. Most of the members belonging to this family are tropical and sub-tropical in distribution. In India the family is represented by many genera, e.g., *Cassia, Bauhinia, Tamarindus, Saraca, Poinciana, Parkinsonia, Caesalpinia,* etc.

- **Habit:** The plants show great variations in their habit, they may be trees or shrubs and very rarely herbs. Mostly they are mesophytes, but xerophytes (e.g., *Parkinsonia*) are also reported. Most of them are wild, but many are cultivated for their beautiful flowers and timber. Some plants are woody climbers. (e.g. *Bauhinia* spp.).

- **Root**: Tap and branched.

- **Stem:** Erect, woody, cylindrical, solid, branched, sometimes herbaceous or climbing. In many species tannin sacs and gum passages are found.

- **Leaves:** The leaves may be simple or compound. If compound they may be pinnate or bi- pinnate. Usually the pinnate leaves are arranged in pairs; petiolate, pulvinus present at the base of the petiole; the pinna ovate or obovate, glabrous, net veined, entire. Usually exstipulate, sometimes minute caducous stipules present.

- **Inflorescence:** Usually racemose, raceme, sometimes pendulous.

- **Flower:** Pedicellate, zygomorphic rarely actinomorphic, hermaphrodite, hypogynous, complete, variously coloured, showy, large or small.

- **Calyx:** Five sepals, free or fused, often petaloid, aestivation imbricate or valvate.

- **Corolla:** Five petals, polypetalous(free); ascending imbricate aestivation, i.e., the posterior petal innermost in bud; inferior, spathulate, showy.

- **Androecium:** Stamens usually ten, free, rarely connate, but, sometimes reduced to staminodes or altogether abortive. In *Cassia* 3 to 5 stamens reduced to staminodes, in *Tamarindus* sp. 3 stamens are well developed, while rest are reduced into staminodes.

- **Gynoecium:** Carpel one (monocarpellary); ovary superior, unilocular; marginal placentation; style long; stigma simple.

- **Fruit:** Pod or lomentum, sessile or stalked, dehiscent or indehiscent.

- **Seed:** Exalbuminous.

- **Pollination:** Anemophilous.

- **Floral Formula:** $\oplus$ or % ☿ K_5 or (5) C_5 A_{10} $\underline{G}_{(1)}$

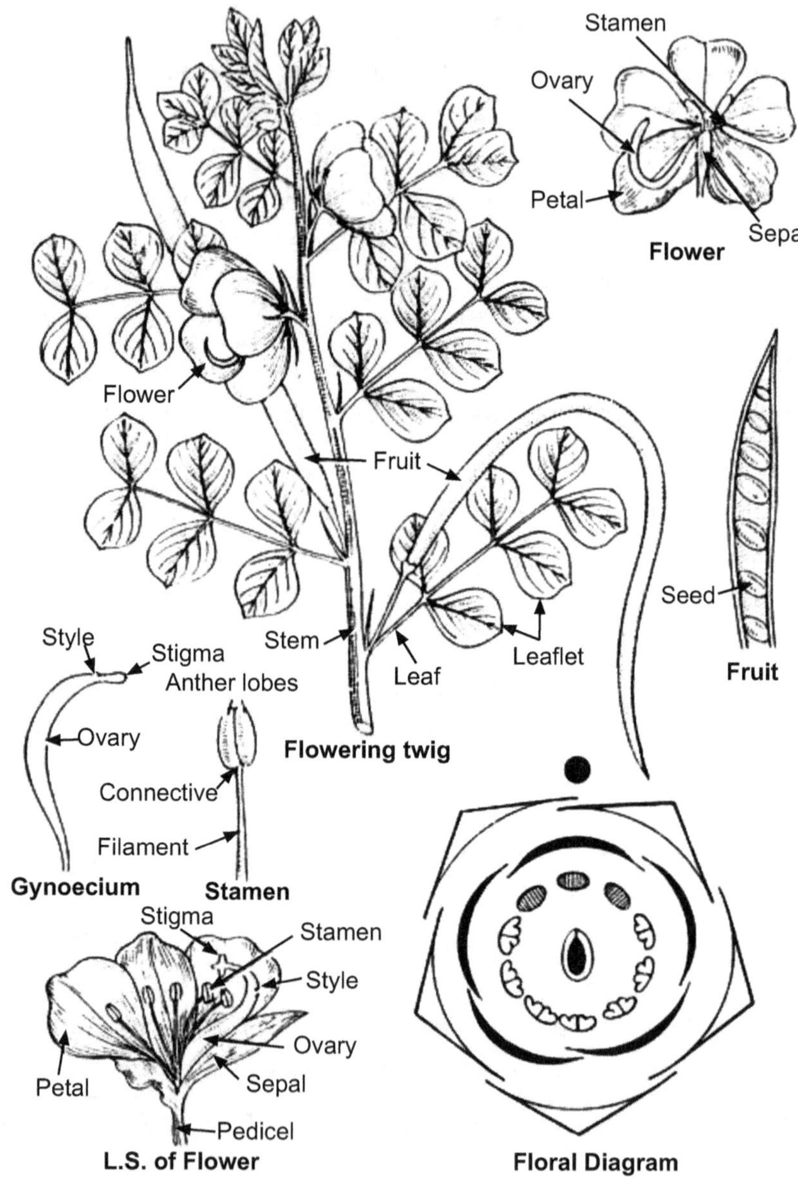

Fig. 5.1: Caesalpiniaceae *Cassia tora* L. Vernacular name - Takala

Distinguishing Characters of the Family:

1. Leaves paripinnate;

2. Flowers zygomorphic;

3. Calyx and corolla 5, ascending imbricate;

4. Stamens 10 or less, free;

5. Gynoecium monocarpellary with marginal placentation.

Economic Importance of Family:

1. **Saraca indica (Sitecha ashok):** This is a tree; native of India. It is grown as an ornamental in the gardens for its scarlet orange flowers. The plant is of great medicinal value.

2. **Cassia fistula (Bahava, Amaltash):** This is a tree, grown as an ornamental. The pulp of the fruits is used as a purgative and laxative.

3. **Cassia auriculata (Tarwad):** The bark is employed for tanning.

4. **Bauhinia racemosa (Aapta):** This is a small bushy tree. The flower buds and fruits are eaten as vegetable. The bark yields a fiber.

5. **Tamarindus indica (Chinch):** This is a tree found throughout our country. It is grown as an avenue. The fruits are edible and are also used as carminative, and laxative. The bark and leaves are employed in tanning. The seeds yield jellose which is used for sizing jute and cotton. The polyose obtained from the seeds, is good substitute for food pectin's.

5.2.2 Solanaceae (*Nightshade* or *Potato Family*)

Systematic position:

Division	: Phanerogams	Flowering plants.
Sub-division	: Angiospermae	Ovules are enclosed in ovary; seeds are enclosed in fruits.

Class	:	Dicotyledons	Seed with 2 cotyledons; flowers pentamerous or tetramerous; leaves with reticulate venation.
Sub-class	:	Gamopetalae	Sepals and petals distinct; petals united.
Series	:	Bicarpellatae	Ovary superior, stamens in one whorl, carpels 2.
Order	:	Polemoniales	Flowers with the stamens adnate to the corolla lobes; a single superior compound ovary.
Family	:	Solanaceae	

- **Distribution:** The family well distributed in tropics and sub-tropics, though a few members are found in temperate zone. The family includes 2,000 species belonging to 90 genera. In India it is represented by 70 species of 21 genera.

- **Habit:** Mostly herbs (Petunia, Withania), shrubs and trees.

- **Root:** A branched tap root system.

- **Stem:** Aerial, erect, climbing, herbaceous, or woody, cylindrical, branched, solid or hollow, hairy, or glabrous, underground stem in *Solanum tuberosum*.

- **Leaves:** Cauline, ramal, exstipulate, petiolate or sessile, alternate sometimes opposite, simple, entire pinnatisect in Lycopersicurn, unicostate reticulate venation.

- **Inflorescence:** Solitary axillary, umbellate cyme, or helicoid cyme in Solanum.

- **Flower:** Bracteate or ebracteate, pedicellate, complete, hermaphrodite, actinomorphic, pentamerous, hypogynous.

- **Calyx:** Sepals 5, gamosepalous, tubular or campanulate, valvate or imbricate, persistent, green or coloured, hairy, inferior.

- **Corolla:** Petals 5, gamopetalous, tubular or infundibuliform, valvate or imbricate aestivation, scale or hair-like outgrowth may arise from the throat of the corolla tube, coloured, inferior.

- **Androecium:** Stamens 5, epipetalous, polyandrous, alternipetalous, filaments inserted deep in the corolla tube, anthers dithecous, usually basifixed or dorsifixed, introrse, inferior.

- **Gynoecium:** Bicarpellary, syncarpous, ovary superior, bilocular, unilocular in Henoonia, axile placentation placentae swollen, many ovules in each loculus, ovary obliquely placed; in some cases nectariferous disc is present; style simple; stigma bifid or capitate.

- **Fruit:** A capsule or beery.

- **Seed:** Endospermic.

- **Pollination:** Entomophilous.

- **Floral formula:** $Br \, \male\female \, \oplus \, K_{(5)} \, C_{(5)} \, A_5 \, \underline{G}_{(4)}$

Distinguishing Characters of the Family:

1. Plants herbs, shurbs rarely trees;

2. Leaves alternate;

3. Inflorescence solitary or in cymes or axillary or terminal;

4. Flowers pentamerous, actinomorphic, hypogynous, hermaphrodite;

5. Calyx persistent, gamosepalous;

6. Corolla gamopetalous, campanulate;

7. Stamens epipetalous;

8. Gynoecium bicarpellary, syncarpous, ovary obliquely placed, axile placentation; placentae swollen; ovules many in each locules;

9. Fruit capsule or berry.

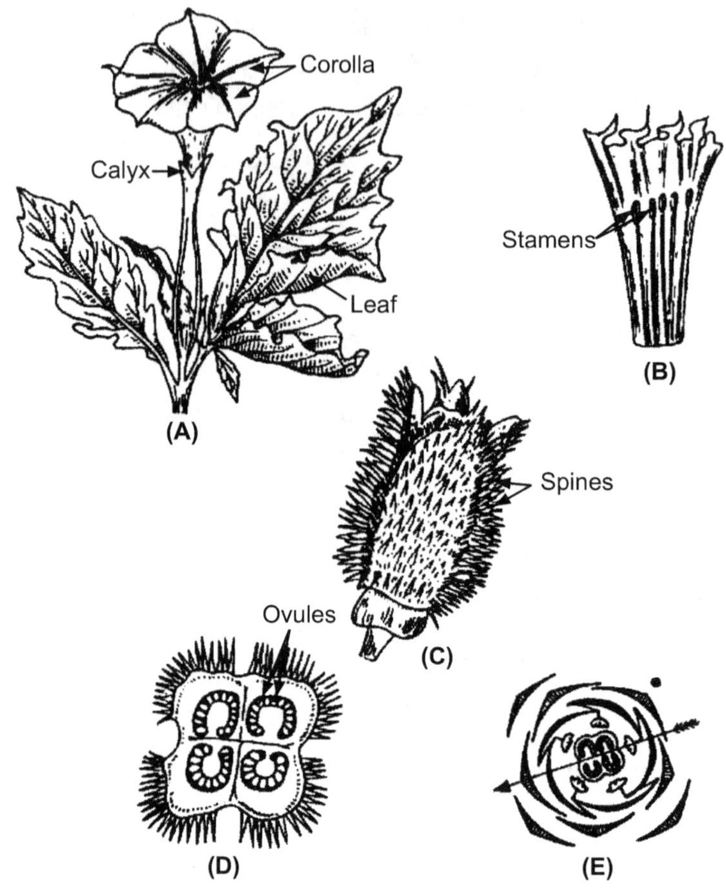

Fig. 5.2: ***Datura stramonium;*** **(A) Habit; (B) Corolla with epipetalous stamens; (C) Capsule; (D) T.S. of ovary (fruit); (E) Floral diagram**

Economic Importance of Family:

1. ***Cestrum nocturnum*** **(Ratrani or Queen of the night):** A garden plant with flowers emitting sweet smell at night.

2. ***Datura metal*** **(Kala Dhotra):** An herb with highly poisonous fruits and seeds. The leaves are used in the treatment of asthma.

3. ***Lycoperscium esculentum*** **(Tomato):** This is cultivated for its edible fruits. This is a common herb. It is native of South America.

4. ***Nicotiana tabacum* (Tambaku):** The leaves are used for smoking and also contain alkoloids, which are used as insecticides. The oil obtained from the seeds is used for burning purposes, and is also used in the manufacture of paints and varnishes. The oil cake makes good manure. The snuffs are prepared from tobacco wastes.

5. ***Withania somnifera* (Ashwangandha):** A perennial tall herb, with potential medicinal value. It contains substances with narcotic and sporific properties. This is a undershrub, found in the drier parts of the country.

5.2.3 Nyctaginaceae (*Four O'clock Family*)

Systematic position:

Division	:	Phanerogams	Flowering plants.
Sub-division	:	Angiospermae	Ovules are enclosed in ovary; seeds are enclosed in fruits.
Class	:	Dicotyledons	Seed with 2 cotyledons; flowers pentamerous or tetramerous; leaves with reticulate venation.
Sub-class	:	Monochlamydeae	Flowers apetalous; perianth lacking or if present not differentiated into sepals and petals.
Series	:	Curvembryeae	embryo coiled, ovule usually 1.
Family	:	Nyctaginaceae	

- **Distribution:** Nyctaginaceae or Four O'clock family includes 30 genera and 300 species according to Rendle. It is widely distributed in tropical and temperate America and warmer parts of the world.

- **Habit:** Herb, shrub or scandent or trees.

- **Root:** Tap root, branched.

- **Stem:** Herbaceous or woody, erect or even climbing e.g. Bougainvillea.

- **Leaf:** Alternate or opposite, simple, those of each pair being often very unequal, exstipulate.

- **Inflorescence:** Cymose, biparous with a tendency to monochasial development in the higher branches.

- **Flower:** Perfect or diclinous as in Pisonia by suppression of stamens or pistil, hypogynous, actinomorphic, rarely zygomorphic in Anisomeris usually subtended by an involucre of 3 to 5 separate or united brightly coloured bracts that are often mistaken for sepals.

- **Perianth:** Tepals 4-5, gamophyllous, tubular with wide spreading lobes, often petaloid (infundibuliform) imbricate or twisted, contorted, connate in a funnel shaped or tubular perigone, the base of which persists and enclosing the fruit forming the so called anthocarp.

- **Androecium:** Stamens variable 2 to 20, free, usually 5 to 8 or equal to number of petals but there may be fewer or more; filaments of unequal lengths.

- **Gynoecium:** One carpel, superior, unilocular with a single basal ovule; style long, simple.

- **Fruit:** Dry, one-seeded anthocarp (achene surrounded by persistent perianth lobes) indehiscent.

- **Seed:** Small, non-endospermic.

- **Pollination:** Entomophilous.

- **Floral formula:** $Br \; \female\male \; \oplus \; P_{(5)} \; A_{5 \text{ or } 2\text{-}20} \; \underline{G}_{(1)}$

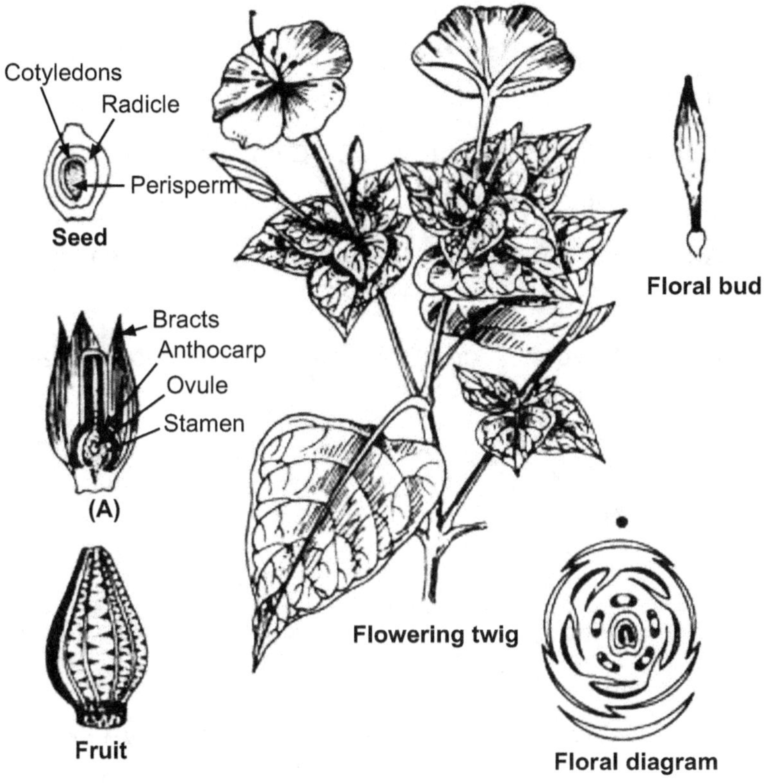

Fig. 5.3: *Mirabilis jalapa*; with different floral part

Distinguishing Characters of the Family:

1. Flowers open in late afternoon;

2. Flowers mostly hermaphrodite rarely diclinous, actinomorphic, usually subtended by an involucre of separate or united brightly coloured bracts or sepaloid bracts that are often mistaken for sepals;

3. Petaloid perianth, tepals 5;

4. Stamens 1-∝;

5. Carpel one.

Economic Importance of Family:

1. ***Mirabilis jalapa*** **(Gulbaksha):** Four O'clock plant H. Gul-e-bas – ornamental plant.

2. ***Boerhaavia diffusa*** **(Punarnava):** Common weed after rainy season, with medicinal goods.

3. ***Bougainvillea spectabilis*** **(Kagadi-Phul):** ornamental shrub with magneta bract.

4. ***Pisonia aculeata*****:** a large climber, armed with recurved axillary spines.

5.2.4 Liliaceae (*Lily Family*)

Systematic position:

Division	: Phanerogams	Flowering plants.
Sub-division	: Angiospermae	Ovules are enclosed in ovary; seeds are enclosed in fruits.
Class	: Monocotyledons	Adventitious root system. Leaves with parallel venation, Flowers trimerous, Cotyledon one.
Series	: Coronarieae	Ovary superior, carpels united, perianth petaloid.
Family	: Liliaceae	

- **Distribution:** 15 genera, 640 species; Widely distributed in the Northern Hemisphere, mainly in the temperate regions.

- **Habitat and Habit:** Mesophytes or xerophytes, mostly herbaceous, persisting by underground tuberous stems. Some perennial herbs are weak-stemmed, climbing, scrambling herbs with leaves modified into hard, curved spines.

- **Roots:** Adventitious root type.

- **Leaves:** Leaves radical, simple alternate. The venation is usually parallel. In some cases leaves are fleshy, in some they are reduced to scales or spines. In some cases the leaf apex and stipules are modified into tendrils.

- **Stem:** Stem is underground tuberous, bulb, corm, and rhizome, aerial; branched, may be herbaceous or woody, solid or fistular, in some cases there are cladodes.

- **Inflorescence:** It is terminal or axillary, solitary or simple raceme on a stout scape. It shows variations like raceme, panicle, cymose, umbel, spike.

- **Flower:** Bracteate, usually hermaphrodite rarely unisexual, actinomorphic, trimerous and hypogynous.

- **Perianth:** The perianths are usually petaloid, arranged into two or three whorls. The tepals are free, equal and regular.

- **Androecium:** Stamens six, epiphyllous, and arranged in two or three whorls.

- **Gynoecium:** Tricarpellary, syncarpous, trilocular with one or more ovules in each locule, axile placentation and superior. The stigma is entire or trilobed.

- **Fruit:** The fruit is globose, berry or coriaceous septicidal capsule or loculicidal capsule. The fruits are many seeded.

- **Pollination:** Entomophilous rarely self-pollination.

- **Floral Formula:** Br, $\oplus$, $\male\female$, $P_{(3+3) \text{ or } 3+3}$, A_{3+3}, $\underline{G}_{(3)}$

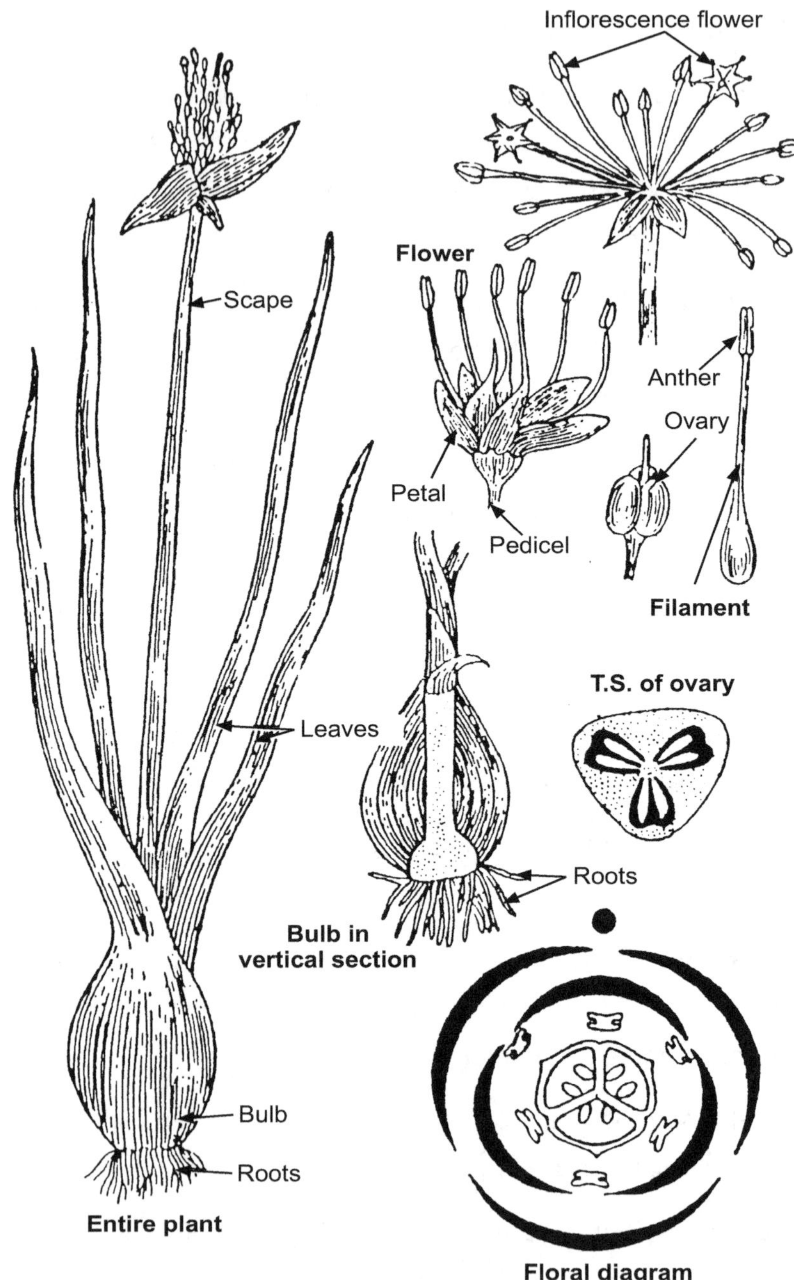

Fig. 5.4: *Allium cepa*

Distinguishing Characters of the Family:

1. Herbs with bulb, corm or rhizome.

2. Leaves alternate or whorled leaves.

3. Flowers bisexual, trimerous

4. Perianth with 6 petaloid tepals.

5. Stamens 6, filaments free.

6. Carpels 3 united, ovary superior, placentation axile.

7. Fruit a capsule.

Economic Importance of Family:

1. ***Aloe vera* (Korphad):** It is one of the constituents of several laxative preparations. It is also valuable in treatment of piles and fissures. The mucilage is useful for inflammations.

2. ***Asparagus officinalis* (Shatavari):** The shoots are consumed as vegetables. Leaves used for decoration in flower arrangement and in bouquet. The roots are used as powerful tonic.

3. ***Chlorophytum tuberosum* (Kusali):** The leaves are consumed as vegetable.

4. ***Dipcadi montanum* (Suichi Bhaji):** Leaves used as vegetables.

5. **Gloriosa superba (Bachnag/Kal-lawi):** An ornamental climbing herb with beautiful flowers. It has many medicinal properties. It is used as a remedy for stomach ache.

6. ***Asphodelus tenuifolius*:** Taken for colds and hemorrhoids (seeds); a febrifuge; used for rheumatic pain.

7. ***Allium cepa* (Onion):** The bulbs are used as vegetable. The green leaves are also edible. The plant posseses medicinal properties and is a stimulant, diuretic and expectorant.

8. ***Allium sativum* (Garlic):** The bulbs are used as a condiment and flavouring substance for vegetables. It also serves as carminative and gastric stimulant in medicinal preparations.

9. ***Asparagus africanus* (African lily):** This is an ornamental herb.

10. ***Pancratium triflorum*:** This is an attractive summer and winter flowering herb, grown as an ornamental plant.

EXERCISE

Q.1   Multiple choice questions: **(01 Mark)**

1. Solanaceae belongs to series..........

   a) Coronarieae

   b) Curvembryeae

   c) Bicarpellatae

   d) Calyciflorae

2. Nyctaginaceae is commonly called as..........

   a) Nightshade or Potato family

   b) Four O'clock family

   c) Lily family

   d) Pumpkin family

3. *Gloriosa superba* belongs to family..........

   a) Solanaceae   b) Liliaceae

   c) Nyctaginaceae   d) Myrtaceae

4. Caesalpiniaceae belongs to sub-class..........

   a) Polypetalae   b) Gamopetalae

   c) Monochlamadae   d) Apetalae

5. Solanaceae belongs to sub-class..........

   a) Polypetalae   b) Gamopetalae

   c) Monochlamadae   d) Apetalae

ANSWERS

1-c	2-b	3-b	4-a	5-b

Q.2 Answer the following: (02 Marks)

1. Enlist the characters of sub-division Angiospermae.

2. Enlist the characters of class Dicotyledonae.

3. Enlist the characters of class Monocotyledonae.

4. Enlist the characters of sub-class: Polypetalae and Gamopetalae.

5. Enlist the characters of orders: Polemoniales & Rosales.

Q.3 Write notes on the following or answer the following:

(03/04 Marks)

1. Write the systematic position of Solanaceae.

2. Write the systematic position of Caesalpiniaceae.

3. Write the systematic position of Nyctaginaceae.

4. Write the systematic position of Liliaceae.

5. Write habit and habitat of Liliaceae.

Q.4 Solve/Short answer of the following: (05 Marks)

1. Enlist the distinguishing characters of Solanaceae.

2. Enlist the distinguishing characters of Caesalpiniaceae.

3. Enlist the distinguishing characters of Nyctaginaceae.

4. Discus about the plants of economic importance in Liliaceae.

5. Discus about the plants of economic importance in Caesalpiiaceae.

Q.5 Answer the following: (08 Marks)

1. Write systematic position, morphological characters and distinguishing characters and economic importance of Caesalpiniaceae.

2. Write systematic position, morphological characters and distinguishing characters and economic importance of Solanaceae.

3. Write systematic position, morphological characters and distinguishing characters and economic importance of Nyctaginaceae.

4. Write systematic position, morphological characters and distinguishing characters and economic importance of Liliaceae.

5. Write systematic position and plants of economic importance of any two families studied by you.

Glossary

(Related to Caesalpiniaceae; Solanaceae; Nyctaginaceae; Liliaceae):

- **Achene:** Small, dry, indehiscent, one-seeded fruit seed separating from ovary wall.

- **Actinomorphic:** A flower having radial symmetry.

- **Adventitious:** Arising from an unusual or irregular position.

- **Alternate:** Arrangement of leaves or parts one at a node, as leaves on a stem. For comparison, opposite and whorled.

- **Axillary:** Borne or carried in the axil.

- **Basifixed:** Fixed to the filament at the base.

- **Beery:** A fleshy, indehiscent, pulpy, multi-seeded fruit resulting from a single pistil, e.g. tomato.

- **Bipinnate:** Twice pinnate, the primary leaflets being again divided into secondary leaflets.

- **Caducous:** Falling off very early as compared to similar structures in other plants.

- **Capsule:** A dry dehiscent fruit produced from a compound pistil, e.g. fruit of a tobacco, *Catalpa*, *Dianthus*.

- **Cauline:** Leaf borne on the main stem.

- **Complete:** The flower with all the four whorls i.e. calyx, corolla, androecium and Gynoecium.

- **Cyme:** A more or less flat-topped determinate inflorescence whose outer flowers open last, e.g. *Sambucus*, elderberry.

- **Dichasial (Biparous) cyme:** Dichotomously branched cymose inflorescence.

- **Dithecous:** Two-celled anther.

- **Dorsifixed:** When filaments appear to be inserted at the back of the anther.

- **Epipetalous:** Stamens borne on the petals or corolla tube *e.g. Justicia, Solanum*

- **Hermaphrodite (bisexual):** The flower having both male and female reproductive organs.

- **Hypogynous:** Situated below the gynoecium or ovary referring to stamens, petals and sepals

- **Imbricate:** Overlapping, as shingles on a roof.

- **Inferior:** Beneath, below; said of an ovary when situated below the apparent point of attachment of stamens and perianth.

- **Introrse:** Facing inward from the centre of flower referred for anthers

- **Monochasial (Uniparous) cyme:** Having a cymose inflorescence with one axis at each branching.

- **Parallel venation-** Veins run parallel to each other Compound inflorescence, stem branches two or more.

- **Persistent:** Remaining attached till maturation

- **Petaloid:** Colored resembling petals

- **Pinnate:** Compound, with leaflets or pinnae arranged feather-like on either side of a common axis or rachis.

- **Polyandrous**: Androecium that consists of free stamens.

- **Ramal:** Leaf borne on the branches.

- **Stipules:** Scale like attachment at the base of the petiole.

- **Syncarpous**: United carpels, compound ovary *e.g. Citrus.*

- **Valvate:** 1. dehiscing by valves; 2. meeting at the edges without overlapping, as leaves or petals in the bud.

- **Venation:** The arrangement of veins in a leaf.

- **Whorl:** Arrangement of three or more structures arising from a single node.

- **Zygomorphic:** Asymmetrical, irregular.